MAMM

River Otter

Moose

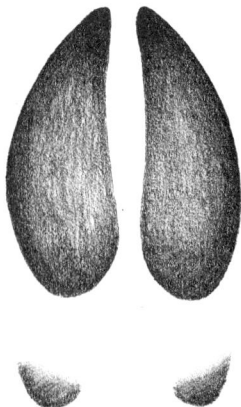

Fore and Hind Prints
Length: 10–18 cm (4–7 in)
Length with dewclaws: to 28 cm (11 in)
Width: 9–15 cm (3.5–6 in)

Straddle
22–50 cm (8.5–20 in)

Stride
Walking: 45–90 cm (1.5–3 ft)
Trotting: to 1.2 m (4 ft)

Size (bull>cow)
Height: 1.5–2 m (5–6.5 ft)
Length: 2.1–2.6 m (7–8.5 ft)

Weight
270–500 kg (600–1100 lb)

walking

MOOSE
Alces alces

The impressive male Moose, the largest of the deer, has a massive rack of antlers. Moose are usually solitary, though you may see a cow with her calf. Despite its placid appearance, a moose may charge humans if approached.

The ungainly shaped Moose moves gracefully, leaving a neat alternating walking pattern. The hind feet direct or double register on the fore prints. Long legs allow for easy movement in snow. Dewclaws—which give extra support for the animal's great weight—register in prints more than 3 cm (1.2 in) deep, but far behind the hoof. In summer, look for tracks in mud beside ponds and other wet areas, where Moose especially like to feed; they are excellent swimmers. In winter, Moose feed in willow flats and coniferous forests, leaving a distinct browseline (highline). Ripped stems and scraped bark, 1.8 m (6 ft) or more above the ground, are additional signs of Moose.

Similar Species: Deer (p. 22) and Caribou (p. 20) tracks are smaller but may be mistaken for a juvenile Moose's.

Caribou

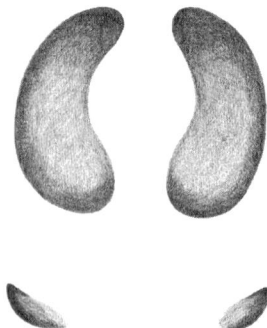

Fore and Hind Prints
Length: 7.6–12 cm (3.0–4.8 in), up
    to 20 cm (8.0 in) with dewclaws
Width: 10–15 cm (4.0–5.8 in)
Straddle
23–36 cm (9.0–14 in)
Stride
Walking: 41–81 cm (16–32 in)
Running: to 1.5 m (5.0 ft)
Group Length: to 2.7 m (9.0 ft)
Size (buck>doe)
Height: 1.1–1.2 m (3.5–4.0 ft)
Length: 2.0–2.6 m (6.5–8.5 ft)
Weight
68–270 kg (150–600 lb)

*walking
(in snow)*

*walking
(on hard
ground)*

CARIBOU
Rangifer tarandus

There are only a few places in Labrador where you might be privileged to see the elegant antlers of this true wilderness animal, which is also called the Barren Ground Caribou. Its sensitivity to human encroachment has reduced its range to parks and extreme northern areas, where it prefers to feed in groups beyond the treeline.

In winter, the Caribou's soft inner sole hardens and shrinks, leaving a firm outer wall that makes neat circles on firmer surfaces. The snowshoe-like hooves spread wide, leaving distinctive large, rounded prints. Big dewclaws help the Caribou to distribute its weight on snow. These dewclaws register behind the forefeet, but rarely the hindfeet; the faster a Caribou runs, the more perpendicular the dewclaws are to the direction of travel. In snow, foot drag is common–look at the print to see how Caribou swing their legs as they walk. Scrape marks indicate where Caribou have dug for lichen hidden beneath the snow.

Similar Species: The Caribou's print size and shape minimize confusion. Juvenile Moose (p. 18) prints may be similar.

21

White-tailed Deer

Fore and Hind Prints
Length: 5–9 cm (2–3.5 in)
Width: 4–6.5 cm (1.6–2.5 in)

Straddle
13–25 cm (5–10 in)

Stride
Walking: 25–50 cm (10–20 in)
Galloping: 1.8–4.5 m (6–15 ft)

Size (buck>doe)
Height: 90–110 cm (3–3.5 ft)
Length: to 1.9 m (6.3 ft)

Weight
55–160 kg (120–350 lb)

walking   *gallop group*

22

WHITE-TAILED DEER
Odocoileus virginianus

The keen hearing of this deer guarantees that it knows about you before you know about it. Frequently, all that we see is its conspicuous white tail in the distance as it gallops away, earning this deer the nickname 'flagtail.' This adaptable deer may be found throughout this region, in small groups at the edges of forests and in brushlands. The White-tailed Deer can be common sights around ranches and in residential areas.

This deer's prints are heart-shaped and pointed. Its alternating walking track pattern shows the hind prints direct registered or double registered on the fore prints. In snow, or when a deer gallops on soft surfaces, the dewclaws register. This flighty deer gallops in the usual style, leaving hind prints ahead of fore prints, with toes spread wide for steadier, safer footing.

Similar Species: Juvenile Moose (p. 18) tracks may be confused with large deer tracks.

Horse

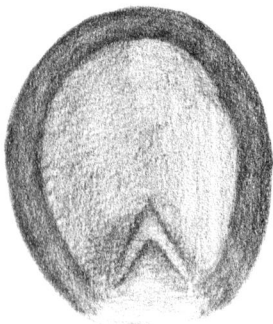

**Fore Print
(hind print is slightly smaller)**
Length: 11–15 cm (4.5–6 in)
Width: 11–14 cm (4.5–5.5 in)

Straddle
5–19 cm (2–7.5 in)

Stride
Walking: 43–70 cm (17–28 in)

Size
Height: to 1.8 m (6 ft)

Weight
to 680 kg (1500 lb)

walking

HORSE
Equus caballus

Back-country use of the popular Horse means that you can expect its tracks to show up almost anywhere.

Unlike any other animal in this book, the Horse has only one huge toe. This toe leaves an oval print with a distinctive 'frog' (V-shaped mark) at its base. When the Horse is shod, the horseshoe shows up clearly as a firm wall at the outside of the print. Not all horses are shod, so do not expect to see this outer wall on every horse print. A typical, unhurried horse trail is an alternating walking pattern, with the hind prints registered on or behind the slightly larger fore prints. Horses are capable of a range of speeds—up to a full gallop—but most recreational horseback riders take a more leisurely outlook on life, preferring to walk their horses and soak up the views!

Similar Species: Mules (rarely shod) have smaller tracks.

Black Bear

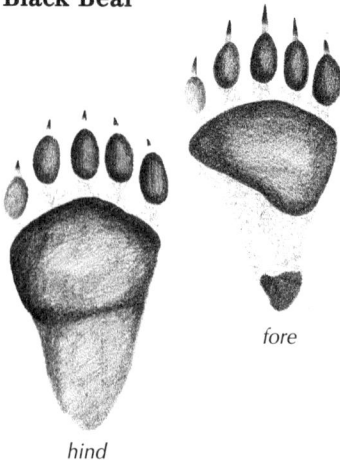

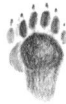

fore

hind

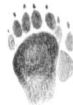

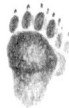

Fore Print
Length: 10–16 cm (4–6.3 in)
Width: 9.5–14 cm (3.8–5.5 in)

Hind Print
Length: 15–18 cm (6–7 in)
Width: 9–14 cm (3.5–5.5 in)

Straddle
23–38 cm (9–15 in)

Stride
Walking: 43–58 cm (17–23 in)

Size (male>female)
Height: 90–110 cm (3–3.5 ft)
Length: 1.5–1.8 m (5–6 ft)

Weight
90–270 kg (200–600 lb)

walking (slow)

BLACK BEAR
Ursus americanus

The Black Bear is scattered in forested areas throughout this region, but do not expect to see its tracks in winter, when it sleeps deeply. Finding fresh bear tracks can be a thrill, but take care—the bear may be just ahead.

Black Bear prints resemble small human prints, but they are wider and show claw marks. The small inner toe rarely registers. The forefoot's small heel pad often shows in the print, and the hind print shows a big heel. The bear's slow walk results in a slightly pigeon-toed double register with the hind print on the fore print. More frequently, at a faster pace, the hind foot oversteps the forefoot. When a bear runs, the two hind feet register in front of the forefeet in an extended cluster. Along well-worn bear paths, look for 'digs'–patches of dug-up earth–and 'bear trees' whose scratched bark shows that these bears climb.

Similar Species: The magnificent Polar Bear has much larger prints but is only found on the northern coast of Newfoundland and Labrador.

Polar Bear

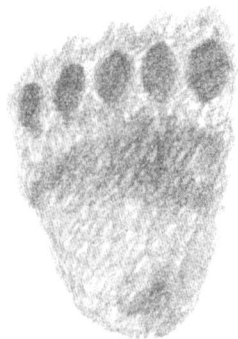

hind

Hind Print
Length: 30–33 cm (12–13 in)
Width: to 23 cm (9.0 in)

Straddle
to 51 cm (20 in)

Size (male>female)
Height: 91–120 cm (3.0–4.0 ft)
Length: 2.0–2.3 m (6.5–7.5 ft)

Weight
270–500 kg (600–1100 lb)

walking (in snow)

POLAR BEAR
Ursus maritimus

The Polar Bear is truly the spirit of the far north. However, few of us will have the pleasure of finding this resilient bear's tracks, let alone the bear itself. Huge and massively furred, the Polar Bear may be found on the far northern coast of Newfoundland and Labrador, where it roams in search of carrion and seals.

The thick fur on a Polar Bear's foot keeps it warm and obscures the finer details of the prints. Clear prints show five toes on each foot; the claws do not always register. When a Polar Bear walks, the hind print registers on or behind the smaller fore print. The Polar Bear's tracks are most likely to be found in snow, where its feet may drag; there is little room for confusion because there is little else in the far north that leaves such enormous tracks.

Similar Species: The smaller Black Bear (p. 26), with similar but smaller prints, can be found in wilderness areas of Atlantic Canada south of the Polar Bear's range.

Coyote

fore

hind

**Fore Print
(hind print is slightly smaller)**
Length: 6–8 cm (2.4–3.2 in
Width: 4–6 cm (1.6–2.4 in)

Straddle
10–18 cm (4–7 in)

Stride
Walking: 20–40 cm (8–16 in)
Trotting: 43–58 cm (17–23 in)
Galloping/Leaping:
    0.8 m–3 m (2.5–10 ft)

Size (female is slightly smaller)
Height: 58–65 cm (23–26 in)
Length: 80–100 cm (32–40 in)

Weight
9–23 kg (20–50 lb)

*walking or
trotting*

gallop group

COYOTE
(Brush Wolf, Prairie Wolf)
Canis latrans

This widespread, adaptable canine prefers open grass-lands or woodlands. It hunts rodents and larger prey, on its own, with a mate or in a family pack. If you find a coyote den—usually a wide-mouthed tunnel leading into a nesting chamber—take care not to bother the family, or the female will have to move her pups to a safer location.

The oval fore prints are slightly larger than the hind prints. The fore heel pad is more triangular than the hind heel pad, which rarely registers clearly. The two outer toes usually do not register claw marks. The Coyote typically walks or trots; the walk has a wider straddle, and the trotting trail is often very straight. When it gallops, the Coyote's hind feet fall in front of its forefeet; the faster it goes, the straighter the gallop group. The Coyote's tail, leaves a dragline in deep snow.

Similar Species: Domestic Dog's (p. 38) less-oval prints splay more, and its trail is erratic. Foot hairs make the smaller Red Fox (p. 34) and Arctic Fox (p. 36) prints less clear.

Gray Wolf

fore

hind

Fore Print
(hind print is slightly smaller)
Length: 10–14 cm (4.0–5.5 in)
Width: 6.4–13 cm (2.5–5.0 in)

Straddle
7.6–18 cm (3.0–7.0 in)

Stride
Walking: 38–81 cm (15–32 in)
Galloping: 91 cm (3.0 ft),
leaps to 2.7 m (9.0 ft)

Size (female is slightly smaller)
Height: 66–97 cm (26–38 in)
Length: 1.1–1.6 m (3.6–5.2 ft)

Weight
32–54 kg (70–120 lb)

walking   *trotting*

GRAY WOLF
(Timber wolf)
Canis lupus

The soulful howl of the wolf epitomizes the outdoor experience, but few people ever hear it—your best bet is national parks or remote, undisturbed areas. The largest of the wild dogs, the wolf may travel in packs or alone, but is rarely seen.

A wolf leaves a straight, alternating track pattern of large oval prints that each show all four claws, with the smaller hind print registering directly on the fore print. The lobing on the fore and hind heel pads differs. In deep snow, wolves sensibly follow their leader, sometimes dragging their feet. When a wolf trots, notice how the hind print has a slight lead and falls to one side, giving an unbalanced appearance. Wolves and Coyotes (p. 30) gallop in the same way.

Similar Species: Domestic Dog (p. 38) prints, rarely as large as a wolf's, fall in a haphazard pattern with a less direct register, and the inner toes tend to splay more. A Wolverine (p. 64) print may show just four toes, but the pad shapes are very different.

Red Fox

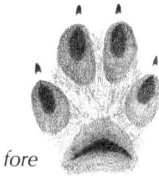

fore

hind

**Fore Print
(hind print is slightly smaller)**
Length: 5.3–7.5 cm (2.1–3 in)
Width: 4–5.8 cm (1.6–2.3 in)

Straddle
5–9 cm (2–3.5 in)

Stride
Trotting: 30–45 cm (12–18 in)
Side-trotting: 36–53 cm (14–21 in)

Size (vixen is slightly smaller)
Height: 35 cm (14 in)
Length: 55–65 cm (22–25 in)

Weight
3.2–7 kg (7–15 lb)

trotting   *side-trotting*

RED FOX
Vulpes vulpes

This beautiful and notoriously cunning fox, found throughout the region, prefers mountainous forests and open areas. It is a very adaptable and intelligent animal.

The finer details of a Red Fox's tracks are blurred by its foot hairs, so only parts of the toes and heel pads show. The horizontal or slightly curved bar across the fore heel pad is diagnostic. A trotting Red Fox leaves a distinctive, straight trail of alternating prints—the hind print direct registers on the wider fore print. When a fox side-trots, it leaves print pairs with the hind print falling to one side of the fore print in typical canid fashion. Foxes gallop like Coyotes (p. 30). The faster the gallop, the straighter the gallop group.

Similar Species: Other canids such as the Arctic Fox (p. 36) lack the bar across the fore heel pad. Domestic Dog (p. 38) prints can be of similar size. Small Coyote prints are similar but have a wider straddle, and the toe marks are more bulbous.

Arctic Fox

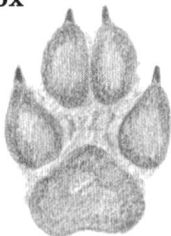

fore

hind

**Fore Print
(hind print slightly smaller)**
Length: 5.1–7.1 cm (2.0–2.8 in)
Width: 5.1–6.4 cm (2.0–2.5 in)

Straddle
5.1–10 cm (2.0–4.0 in)

Stride
Walking/Trotting: 18–30 cm
(7.0–12 in)

Size
Height: to 30 cm (12 in)
Length with tail: 76–91 cm
(30–36 in)

Weight
2.5–4.0 kg (5.5–8.8 lb)

loping

running

ARCTIC FOX
Alopex lagopus

    This delightful fox of the north changes from having a thick and warm, white winter coat to a thinner, blue-brown one in the summer. Although it is a solitary animal, Arctic Fox can be curious; in remote areas of Labrador it may come quite close to observe you.

    Arctic Fox tracks are typical of dog family tracks, with the hind prints slightly smaller than the fore prints. Both fore and hind prints show four toes, and the claws usually register. The close-set toes register quite clearly in summer. In winter, however, when thick fur covers the pads for warmth, print detail is obscured. When this fox walks or trots, it leaves a neat trail of alternating prints nearly in a line. If you come across an old bear or wolf kill, look around for fox tracks, because foxes frequently follow other carnivores to feed on their leftovers.

Similar Species: The Red Fox (p. 34), with slightly smaller prints, has a distinctive heel-pad bar, but this feature may be obscured by winter hair.

Domestic Dog

fore

hind

Fore Print
(hind print is smaller)
Length: 2.5–14 cm (1–5.5 in)
Width: 2.5–13 cm (1–5 in)
Straddle
3.8–20 cm (1.5–8 in)
Stride
Walking: 7.5–80 cm (3-32 in)
Loping to Galloping: to 2.7 m (9 ft)
Size
Very variable
Weight
Very variable

slow trotting

loping to galloping

DOMESTIC DOG

Canis familiaris

Dogs come in many shapes and sizes—from the tiny Chihuahua with its dainty feet to the robust and powerful Great Dane. Consequently, Domestic Dog tracks vary enormously. Dog ownership is high in many residential areas, and the popular pastime of dog walking can result in many dog tracks being left scattered about, especially if there is wet mud or snow.

The forefeet of the Domestic Dog, which are much larger than the hindfeet and support more of the animal's weight, leave the clearest tracks. When a dog walks, the hind prints usually register ahead of or beside the fore prints. As the dog moves faster, it trots and then lopes before it gallops. In a trot or lope pattern the prints alternate fore-hind-fore-hind, whereas a gallop group shows (from back to front) fore-fore-hind-hind.

Similar Species: Keep in mind that dog tracks are usually found close to human tracks or activity. Grey Wolf (p. 32), with large dog-like tracks, prefers wilderness areas. Fox (pp. 34-37) tracks may be confused with small dog tracks.

Lynx

fore

hind

Fore Print
(hind print is slightly smaller)
Length: 9–11 cm (3.5–4.5 in)
Width: 9–12 cm (3.5–4.8 in)

Straddle
15–23 cm (6–9 in)

Stride
Walking: 30–70 cm (12–28 in)

Size
Length: 75–90 cm (2.5–3 ft)

Weight
7–14 kg (15–30 lb)

walking

LYNX
Lynx canadensis

This large cat is a thrill to see, but it is sensitive to human interference. The elusive Lynx is abundant only in remote, undisturbed, dense forests. With huge feet and a relatively lightweight body, it stays on top of the snow as it pursues its main prey, the Snowshoe Hare (p. 68).

This cautious walker leaves a neat alternating track pattern, the hind print direct registered on top of the fore print. Thick fur on the feet often results in prints that are big, round depressions with no detail. In deeper snow, the print may be extended by 'handles' off to the rear. However deep the snow, this cat sinks no more than 20 cm (8 in), and it rarely drags its feet. The Lynx is more likely to bound than to run. Its curious nature results in a meandering trail that may lead to a partially buried food cache.

Similar Species: Bobcat (p. 42) prints are smaller, and the trail may show draglines. Fisher (p. 52) prints may not show the fifth toe—look for mustelid habits. Canid prints (pp. 30-39) show claw marks, their length exceeds their width, and the front of the footpad has one lobe.

Bobcat

fore

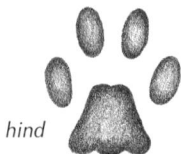

hind

Fore Print
(hind print is slightly smaller)
Length: 4.5–6.5 cm (1.8–2.5 in)
Width: 4.5–6.5 cm (1.8–2.5 in)

Straddle
10–18 cm (4–7 in)

Stride
Walking: 20–40 cm (8–16 in)
Running: 1.2–2.4 m (4–8 ft)

Size
(female is slightly smaller)
Height: 50–55 cm (20–22 in)
Length: 65–75 cm (25–30 in)

Weight
7–16 kg (15–35 lb)

walking

*ambling
to loping*

BOBCAT (Wildcat)

Lynx rufus

The widely distributed Bobcat, a stealthy and usually nocturnal hunter, is seldom seen. Very adaptable, it can leave tracks anywhere from wild mountainsides to chaparral and even into residential areas.

When the Bobcat walks, its hind feet usually register directly on the larger fore prints. As the Bobcat picks up speed, its trail becomes an ambling pattern of paired prints, the hind leading the fore. At even greater speeds, it leaves four-print groups in a lope pattern. Especially the fore prints show asymmetry. The front part of the heel pad has two lobes and the rear part has three. In deep snow the Bobcat's feet leave draglines. The Bobcat marks its territory with half-buried scat along its meandering trail.

Similar Species: Large Domestic Cats (p. 44) have similar prints but a shorter stride and narrower straddle. Canid (pp. 30-39) prints are narrower than they are long and show claw marks, and the fronts of their footpads are once-lobed; wild canid trails do not meander. Some mustelids' (pp. 50–65) hind prints may also seem similar—look for mustelid habits.

Domestic Cat

fore

hind

Fore Print
(hind print is slightly smaller)
Length: 2.5–4 cm (1–1.6 in)
Width: 2.5–4.5 cm (1–1.8 in)

Straddle
6–11 cm (2.4–4.5 in)

Stride
Walking: 13–20 cm (5–8 in)
Loping/Galloping:
    35–80 cm (14–32 in)

Size (male>female)
Height: 50–55 cm (20–22 in)
Length with tail: 75 cm (30 in)

Weight
3–6 kg (6.5–13 lb)

walking

loping to
galloping

DOMESTIC CAT
(House Cat)
Felis catus

The tracks of the familiar and abundant Domestic Cat can show up almost any place where there are people. Abandoned cats may roam farther afield, and these 'feral cats' lead a pretty wild and independent existence. Domestic Cats can come in many shapes, sizes and colours.

As with all felines, a Domestic Cat's fore print and slightly smaller hind print both show four toe pads. Its retractable claws, kept clean and sharp for catching prey, do not register. Cat prints usually show a slight asymmetry, with one toe leading the others. A Domestic Cat makes a neat alternating walking track pattern, usually in direct register, as one would expect from this animal's fastidious nature. When a cat picks up speed, it leaves clusters of four prints, the hind feet registering in front of the forefeet.

Similar Species: A small Bobcat (p. 42) may leave tracks similar to a very large Domestic Cat's. Red Fox (p. 34) and Domestic Dog (p. 38) prints show claw marks.

Raccoon

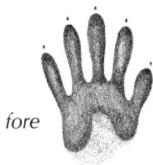

fore

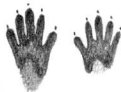

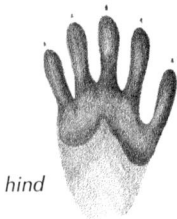

hind

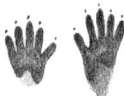

Fore Print
Length: 5–7.5 cm (2–3 in)
Width: 4.5–6.5 cm (1.8–2.5 in)

Hind Print
Length: 6–9.5 cm (2.4–3.8 in)
Width: 5–6.5 cm (2–2.5 in)

Straddle
8.5–15 cm (3.3–6 in)

Stride
Walking: 20–45 cm (8–18 in)
Bounding: 38–65 cm (15–25 in)

Size (female is slightly smaller)
Length: 60–95 cm (24–37 in)

Weight
5–16 kg (11–35 lb)

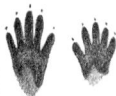

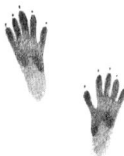

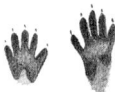

walking

bounding group

RACCOON
Procyon lotor

The inquisitive Raccoon is common in the region. It is adored by some people for its distinctive face mask, yet is disliked for its boundless curiosity—often expressed with residential garbage cans. A good place to look for its tracks is near water at low elevations. The Raccoon likes to rest in trees. It usually dens up for the colder months.

The Raccoon's unusual print, showing five well-formed toes, looks like a human handprint; its small claws make dots. Its highly dexterous forefeet rarely leave heel prints, but its hind prints, which are generally much clearer, do show heels. The Raccoon's peculiar walking track pattern shows the left fore print next to the right hind print (or just in front) and vice versa. On the rare occasions when a Raccoon is out in deep snow, it may use a direct-registering walk. The Raccoon occasionally bounds, leaving clusters with the hind prints in front of the fore prints.

Similar Species: Raccoon prints are normally distinctive, but, in deep snow, Fisher (p. 52) or River Otter (p. 50) tracks may look similar; so can Woodchuck (p. 76) tracks.

Harbour Seal

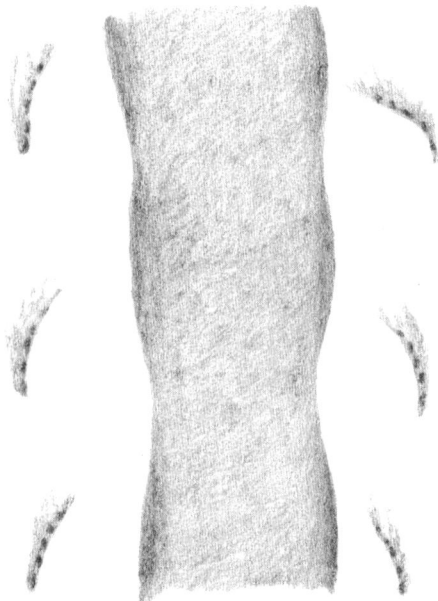

beach track

Size (male>female)
Length: 1.2–1.8 m (4–6 ft)
Weight
80–140 kg (180–310 lb)

HARBOUR SEAL
Phoca vitulina

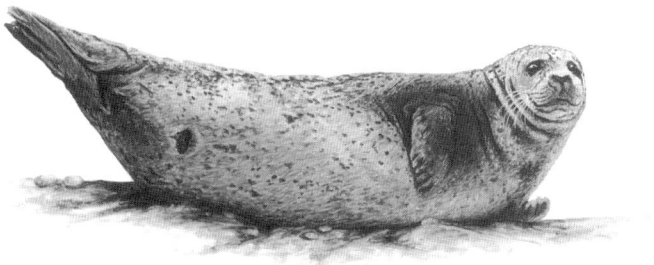

This common seal frequents isolated coastal beaches. One of the smaller seals, it is quite shy and will usually slide off its rocky sentry post into the sea to escape naturalists. Look in sandy and muddy areas between platforms for its tracks—unmistakable because of their location and large size. Sometimes a Harbour Seal works its way up a river, so you may find its tracks along riverbanks.

Seals, though unrivalled in the water, are not the most graceful of animals on land. A seal's heavy, fat body and flipper-like feet can leave messy tracks: a wide trough (made by its cumbersome belly) with dents alongside (made as the seal pushed itself along with its forefeet). Look for the marks made by the seal's nails.

Similar Species: The Grey Seal (*Halichoerus grypus*), more than twice as large as the little Harbour Seal, may be found on the Maritimes' rocky (sometimes sandy) coasts.

River Otter

fore

hind

Fore Print
Length: 6.5–9 cm (2.5–3.5 in)
Width: 5–7.5 cm (2–3 in)
Hind Print
Length: 7.5–10 cm (3–4 in)
Width: 5.8–8.5 cm (2.3–3.3 in)
Straddle
10–23 cm (4–9 in)
Stride
Loping: 30–70 cm (12–27 in)
Size
(female is two-thirds the size of male)
Length with tail: 90–130 cm (3–4.3 ft)
Weight
4.5–11 kg (10–25 lb)

loping (fast)

RIVER OTTER
Lontra canadensis

No animal knows how to have more fun than a River Otter. If you are lucky enough to watch one at play, you will not soon forget the experience. Widespread and well-adapted for the aquatic environment, this otter lives along waterbodies; an otter in the forest is usually on its way to another waterbody. Expect to see a wealth of evidence of an otter's presence along the riverbanks in its home territory.

In soft mud, the River Otter's five-toed feet, especially the hind ones, register evidence of webbing. The inner toes are set slightly apart. If the forefoot's metacarpal pad registers, it lengthens the print. Very variable, otter trails usually show a typical mustelid 2×2 loping. However, with faster gaits they can sometimes show groups of four and three prints. The thick, heavy tail often leaves a dragline. This otter loves to slide in snow, often down riverbanks, leaving troughs nearly 30 cm (1 ft) wide. In summer it rolls and slides on the grass and in the mud.

Similar Species: Other mustelid trails lack conspicuous tail drag. The Marten (p. 54) has similar-sized prints. Mink (p. 56) prints are about half the size. The Fisher (p. 52) usually has hairy feet, with forefeet larger than hind feet.

Fisher

fore

Fore Print
Length: 5.3–10 cm (2.1–4 in)
Width: 5.3–8.5 cm (2.1–3.3 in)
Hind Print
Length: 5.3–7.5 cm (2.1–3 in)
Width: 5–7.5 cm (2–3 in)
Straddle
7.5–18 cm (3–7 in)
Stride
Walking: 18–35 cm (7–14 in)
2×2 loping: 30–130 cm (1–4.3 ft)
Loping: 30–90 cm (1–3 ft)
Size (male>female)
Length with tail:
    85–100 cm (34–40 in)
Weight
1.4–5.5 kg (3–12 lb)

walking   *2×2 loping*

FISHER (Black Cat)

Martes pennanti

This agile hunter is comfortable
both on the ground and
in the trees of
mixed hard-
wood forests.
The Fisher's speed and eager hunting
antics make for exciting tracking as it races up trees and
along the ground in its quest for squirrels. It is one of the
few predators to kill and eat Porcupines (p. 70).

Though five toes may register, the small inner toe
frequently does not. Only the forefoot has a small heel pad
that can show up in the print. The Fisher occasionally
walks, making a direct-registering alternating track pattern,
but it more often 2×2 lopes in typical mustelid fashion,
leaving angled print pairs with the hind print direct registered on the fore print. Loping, its most common gait, pro-
duces three- and four- print groups (see the River Otter,
p. 50). The patterns often vary within a short distance. Not
associated with water, Fishers have been misnamed!

Similar Species: Male Marten (p. 54) tracks may be
confused with a small female Fisher's, but Martens weigh
less and leave shallower prints. Otters have larger hind feet
than forefeet. When only four toes register, Fisher prints
may look like a Bobcat's (p. 42).

Marten

Fore and Hind Prints
Length: 4.5–6.5 cm (1.8–2.5 in)
Width: 3.8–7 cm (1.5–2.8 in)
Straddle
6.5–10 cm (2.5–4 in)
Stride
Walking: 10–23 cm (4–9 in)
2×2 loping: 23–120 cm (9–46 in)
Size (male>female)
Length with tail:
    53–70 cm (21–28 in)
Weight
0.7–1.3 kg (1.5–2.8 lb)

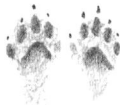

walking   *2×2 loping*

MARTEN
(American Sable)
Martes americana

This aggressive predator is found in the coniferous and mixed-wood forests of much of Atlantic Canada, except Labrador.

The Marten seldom leaves a clear print—often just four toes register and the heel pad is undeveloped. In winter the hairiness of the feet often blurs all pad detail, especially the poorly developed palm pads. In the Marten's alternating walk, the hind foot registers on the fore print. In 2×2 loping, the hind prints fall on the fore prints to form slightly angled print pairs in a typical mustelid pattern. Its loping track patterns may appear as three- or four-print clusters (see the River Otter, p. 50). Follow the criss-crossing trails— if a Marten has scrambled up a tree, look for a sitzmark where it has jumped down.

Similar Species: Size and habitat are often key to distinguishing Marten, Fisher (p. 52) and Mink (p. 56) tracks. Female Fisher prints may resemble a large male Marten's, but they will be clearer. Male Mink prints overlap in size with small female Marten prints, but Mink rarely climb trees and (unlike Martens) are usually found near water.

Mink

fore

hind

Fore and Hind Prints
Length: 3.3–5 cm (1.3–2 in)
Width: 3.3–4.5 cm (1.3–1.8 in)

Straddle
5.3–9 cm (2.1–3.5 in)

Stride
Walking/2×2 loping: 20–90 cm (8–36 in)

Size (male>female)
Length with tail: 48–70 cm (19–28 in)

Weight
0.7–1.6 kg (1.5–3.5 lb)

2×2 loping

MINK
Mustela vison

The lustrous Mink, which is widespread throughout Atlantic Canada, prefers watery habitats surrounded by brush or forest. At home as much on land as in water, this nocturnal hunter can be exciting to track. Like the River Otter (p. 50), the Mink sometimes slides in snow, carving out a trough up to 15 cm (6 in) wide for an observant tracker to spot.

The Mink's fore print shows five (perhaps four) toes, with five loosely connected palm pads in an arc, but the hind print shows only four palm pads. The metacarpal pad of the forefoot rarely registers, but the furred heel of the hind foot may register, lengthening the hind print. The Mink prefers the typical mustelid 2×2 loping, making consistently spaced, slightly angled double prints. Its diverse track patterns also include alternating walking; loping with three- and four-print groups (like the River Otter); and bounding (like a Snowshoe Hare, p. 68).

Similar Species: Small Martens (p. 54) may have similar prints, but without a consistent 2×2 loping gait, and they do not live near water. Weasels (pp. 58–63) make similar but smaller tracks.

Least Weasel

Fore and Hind Prints
Length: 1.3–2.0 cm (0.5–0.8 in)
Width: 1.0–1.3 cm (0.4–0.5 in)

Straddle
2.0–3.8 cm (0.8–1.5 in)

Stride
2x2 loping: 13–50 cm (5.0–20 in)

Size (male>female)
Length with tail: 17–23 cm
(6.5–9.0 in)

Weight
37–65 g (1.3–2.3 oz)

2x2 loping

LEAST WEASEL
Mustela nivalis

The typical weasel gait is a 2x2 lope, leaving a trail of paired prints. Because of a weasel's light weight and small, hairy feet, the pad detail is often unclear, especially in snow. Even with clear tracks, the inner (fifth) toe rarely registers. To identify the weasel species, pay close attention to straddle and stride, but note that small females of a larger species and large males of a smaller species may have similar tracks. Also check the habit displayed in loping patterns and note the distribution and habitat.

This weasel, found throughout Labrador, is the smallest weasel, and has the least-clear tracks. Tracks may be found around wetlands and in open woodlands and fields.

Similar Species: A large male's tracks may resemble those of a small female Short-tailed Weasel (p. 60), but the latter does not frequent wet areas, preferring upland areas and dry woodlands.

Short-tailed Weasel

Fore and Hind Prints
Length: 2–3.3 cm (0.8–1.3 in)
Width: 1.3–1.5 cm (0.5–0.6 in)
Straddle
2.5–5.3 cm (1–2.1 in)
Stride
2×2 loping: 23–90 cm (9–36 in)
Size (male>female)
Length with tail 20–35 cm: (8–14 in)
Weight
30–170 g (1–6 oz)

2×2 loping

SHORT-TAILED WEASEL
(Ermine, Stoat)
Mustela erminea

The Short-tailed Weasel is smaller than the Long-tailed Weasel (p. 62) and is more widely distributed. It prefers woodlands and meadows up to higher elevations but does not favour wetlands or dense coniferous forests.

The usual gait for any weasel is a 2×2 lope that results in a trail of paired prints. A weasel's light weight and small, hairy feet mean that the pad detail is often unclear, especially in snow. Even with clear tracks, the inner (fifth) toe rarely registers. This weasel's 2×2 loping tracks may fall in clusters, with alternating short and long strides.

Similar Species: Small female Long-tailed Weasel tracks may be the same size as a large male Short-tailed Weasel's. Large Least Weasel (p. 58) tracks may be similar to small Short-tailed Weasel tracks.

Long-tailed Weasel

Fore and Hind Prints
Length: 2.8–4.5 cm (1.1–1.8 in)
Width: 2–2.5 cm (0.8–1 in)

Straddle
4.5–7 cm (1.8–2.8 in)

Stride
2×2 loping: 24–110 cm (9.5–43 in)

Size (male>female)
Length with tail:
    30–55 cm (12–22 in)

Weight
85–340 g (3–12 oz)

2×2 loping

LONG-TAILED WEASEL
Mustela frenata

On a sunny winter day, there may be no better wildlife experience than to follow the tracks of a Long-tailed Weasel. This curious animal zigzags as though it cannot make up its mind which way to go, and every object it encounters seems to offer a monetary distraction. These weasels have high energy, a characteristic easily read in the tracks left as it leaps, bounds and circles through its territory.

This weasel is larger than a Short-tailed Weasel, and it has a smaller distribution in Atlantic Canada. Its typical gait is a 2x2 lope, with an irregular stride length–sometimes short and sometimes long–and no consistent behaviour. Even with clear tracks, the inner (fifth) toe rarely registers. In winter, look for holes where the weasel has suddenly plunged into the deep snow, perhaps in pursuit of prey.

Similar Species: Large male Short-tailed Weasel (p. 60) prints overlap in size with small female Long-tailed weasels.

Wolverine

fore

hind

Fore Print
Length with heel: 10–19 cm
(4.0–7.5 in)
Width: 10–13 cm (4.0–5.0 in)
Hind Print
Length: 8.9–10 cm (3.5–4.0 in)
Width: 10–13 cm (4.0–5.0 in)
Straddle
18–23 cm (7.0–9.0 in)
Stride
Walking: 7.6–30 cm (3.0–12 in)
Running: 25–102 cm (10–40 in)
Size (female slightly smaller)
Height: 41 cm (16 in)
Length: 81–117 cm (32–46 in)
Weight
8.1–21 kg (18–47 lb)

loping (slow)

WOLVERINE
Gulo gulo

The reputation of the robust and powerful Wolverine, largest of the mustelids, has earned it many nicknames, such as 'skunk bear' and 'Indian devil.' Wolverines live in coniferous forests, and their need for pristine wilderness has resulted in a scarce and scattered distribution throughout much of their range.

As with other mustelids, Wolverines have five toes, but the inner (fifth) toe rarely registers in the print. Though the small heel pad of the forefoot usually registers, the heel is rarely registered on the hind print. Because of its low, squat shape, the Wolverine leaves a host of erratic but typical mustelid trails: an alternating walking pattern, the typical 2x2 loping pattern with its print pairs, and the common loping pattern of three- and four-print groups.

Similar Species: Wolverine tracks are more erratic and much larger than those of other mustelids. An unclear or partial print may resemble those of canids (pp. 30-39).

Striped Skunk

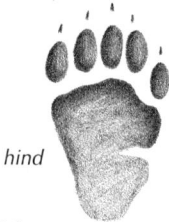

fore

hind

Fore Print
Length: 3.8–5.6 cm (1.5–2.2 in)
Width: 2.5–3.8 cm (1–1.5 in)
Hind Print
Length: 3.8–6.5 cm (1.5–2.5 in)
Width: 2.5–3.8 cm (1–1.5 in)
Straddle
 7–11 cm (2.8–4.5 in)
Stride
Walking/Bounding:
    6.5–20 cm (2.5–8 in)
Size
Length with tail:
    50–80 cm (20–32 in)
Weight
 2.7–6.5 kg (6–14 lb)

walking (fast)   *running*

STRIPED SKUNK
Mephitis mephitis

This striking skunk has a notorious reputation for its vile smell, and the lingering odour is often the best sign of its presence. Widespread throughout Atlantic Canada in a diversity of habitats, it prefers lower elevations. The Striped Skunk dens up in winter, coming out on warmer days and in spring.

Forefeet and hind feet each have five toes. The long claws on the forefeet often register. The smooth palm pads and small heel pads leave surprisingly small prints. Skunks mostly walk—with such a potent smell for their defence, and those memorable black and white stripes, they rarely need to run. Note that the skunk's trail rarely shows any consistent pattern, though an alternating walking pattern may be evident. The greater a skunk's speed, the more the hind foot oversteps the fore. If it runs, its trail consists of clumsy, closely set four-print groups. In snow it drags its feet.

Similar Species: There are no other wild skunks in Atlantic Canada. Mustelid (pp. 50–65) tracks are farther apart than those resulting from a skunk's shuffling gait, and skunk prints do not overlap.

Snowshoe Hare

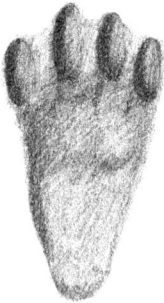

hind

fore

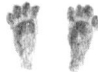

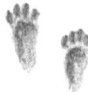

hopping

Fore Print
Length: 5–7.5 cm (2–3 in)
Width: 3.8–5 cm (1.5–2 in)

Hind Print
Length: 10–15 cm (4–6 in)
Width: 5–9 cm (2–3.5 in)

Straddle
15–20 cm (6–8 in)

Stride
Hopping: 25–130 cm (0.8–4.3 ft)

Size
Length: 30–53 cm (12–21 in)

Weight
0.9–1.8 kg (2–4 lb)

SNOWSHOE HARE
(Varying Hare)
Lepus americanus

This hare is well known
for its colour change from
summer brown to winter
white and for its huge
hind feet, which
enable it to 'float'
on top of snow.
Widespread throughout this region, it frequents brushy
areas in forests, which provide good cover from the Lynx
(p. 40) and the Coyote (p. 30), its most likely predators.
Hares are most active at night.

The Snowshoe Hare's most common track pattern is a
hopping one, with triangular four-print groups; they can be
quite long if the hare moves quickly. In winter, heavy fur
on the hind feet (much larger than the forefeet) thickens the
toes, which can splay out to further distribute the hare's
weight on snow. Hares make well-worn runways that are
often used as escape runs. You may encounter a resting
hare, because hares do not live in burrows. Twigs and stems
neatly cut at a 45° angle also indicate this hare's presence.

Similar Species: The Arctic Hare *(Lepus arcticus)* makes
similar tracks but lives only in Newfoundland and northern
Labrador.

Porcupine

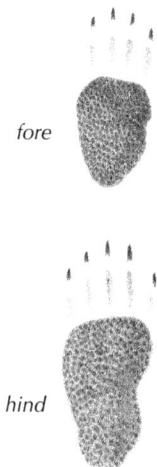

fore

hind

Fore Print
Length: 5.8–8.5 cm (2.3–3.3 in)
Width: 3.3–4.8 cm (1.3–1.9 in)
Hind Print
Length: 7–10 cm (2.8–4 in)
Width: 3.8–5 cm (1.5–2 in)
Straddle
14–23 cm (5.5–9 in)
Stride
Walking: 13–25 cm (5–10 in)
Size
Length with tail: 65–100 cm (25–40 in)
Weight
4.5–13 kg (10–28 lb)

walking

PORCUPINE
Erethizon dorsatum

This notorious rodent rarely runs—its many long quills are a formidable defence. Widespread in this region except in Newfoundland, the Porcupine prefers forests, but it can also be seen in more open areas.

The Porcupine's preferred pigeon-toed, waddling gait leaves an alternating track pattern, with the hind print registered on or slightly in front of the shorter fore print. Look for long claw marks on all prints. The fore print shows four toes, and the hind print shows five. On clear prints, the unusual pebbly surface of the solid heel pads may show, but a Porcupine's tracks are often scratch-marked by its heavy, spiny tail. In deeper snow this squat animal drags its feet, and it may leave a trough with its body. A Porcupine's trail might lead you to a tree, where this animal spends much of its time feeding; if so, look for chewed bark or nipped twigs on the ground.

Similar Species: The Porcupine's prints are different than those of any other similar-sized animal in this region.

Beaver

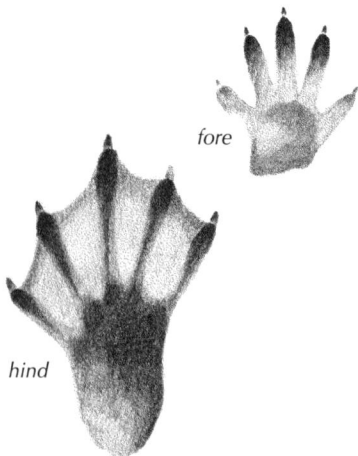

fore

hind

Fore Print
Length: 6.5–10 cm (2.5–4 in)
Width: 5–9 cm (2–3.5 in)

Hind Print
Length: 13–18 cm (5–7 in)
Width: 8.5–13 cm (3.3–5.3 in)

Straddle
15–28 cm (6–11 in)

Stride
Walking: 7.5–17 cm (3–6.5 in)

Size
Length with tail: 90–120 cm (3–4 ft)

Weight
13–34 kg (28–75 lb)

walking

BEAVER
Castor canadensis

Few animals leave as many signs of their presence as the Beaver, the largest North American rodent and a common sight around water. Look for the conspicuous dams and lodges—capable of changing the local landscape—and the stumps of felled trees. Check trunks gnawed clean of bark for marks of the Beaver's huge incisors. A scent mound marked with castoreum, a strong-smelling yellowish fluid that Beavers produce, also indicates recent activity.

The Beaver's thick, scaly tail may mar its tracks, as can the branches that it drags about for construction and food. Check the large hind prints for signs of webbing and broad toenails–the nail of the second inner toe usually does not show. Rarely do all five toes on each foot register. Irregular foot placement in the alternating walking gait may produce a direct register or a double register. Repeated path use results in well-worn trails.

Similar Species: The Beaver's many signs, including its large hind prints, minimize confusion. Muskrat (p. 74) prints are smaller.

Muskrat

fore

hind

Fore Print
Length: 2.8–3.8 cm (1.1–1.5 in)
Width: 2.8–3.8 cm (1.1–1.5 in)

Hind Print
Length: 4–8 cm (1.6–3.2 in)
Width: 3.8–5.3 cm (1.5–2.1 in)

Straddle
7.5–13 cm (3–5 in)

Stride
Walking: 7.5–13 cm (3–5 in)
Running: to 30 cm (1 ft)

Size
Length with tail: 40–65 cm (16–25 in)

Weight
0.9–1.8 kg (2–4 lb)

walking

MUSKRAT
Ondatra zibethicus

Like the Beaver (p. 72), this rodent is found throughout Atlantic Canada, wherever there is water. Beavers are very tolerant of Muskrats and even allow them to live in parts of their lodges. Active year-round, the Muskrat leaves plenty of signs. It digs extensive networks of burrows, often undermining riverbanks, so do not be surprised if you suddenly fall into a hidden hole! Also look for small lodges in the water and beds of vegetation on which the Muskrat rests, suns and feeds in summer.

The small inner toe of the five on each forefoot rarely registers. The hind print shows five well-formed toes that may have a 'shelf' around them, created by stiff hairs that aid in swimming. The common alternating walking pattern shows print pairs that alternate from side to side, with the hind print just behind or slightly overlapping the fore print. In snow, a Muskrat's feet drag, and its tail leaves a sweeping dragline in its tracks.

Similar Species: Few animals share this water-loving rodent's habits. The Beaver makes larger tracks and leaves many other signs.

Woodchuck

fore

hind

Fore and Hind Prints
Length: 4.5–7 cm (1.8–2.8 in)
Width: 2.5–5 cm (1–2 in)
Straddle
8.5–15 cm (3.3–6 in)
Stride
Walking: 5–15 cm (2–6 in)
Running: 15–35 cm (6–14 in)
Size (male>female)
Length with tail:
    50–65 cm (20–25 in)
Weight
2.5–5.5 kg (5.5–12 lb)

walking   *bounding*

WOODCHUCK
(Whistle Pig, Groundhog, Marmot)
Marmota monax

This robust member of the squirrel family is a common sight in open woodlands and adjacent open areas throughout the region, excluding Newfoundland. Always on the watch for predators, but not too troubled by humans, the Woodchuck never wanders far from its burrow. This marmot hibernates during winter but emerges in early spring; look for tracks in late spring snowfalls and in the mud around burrow entrances.

A Woodchuck's fore print shows four toes, three palm pads and two heel pads (not always evident). The hind print shows five toes, four palm pads and two poorly registering heel pads. The Woodchuck usually leaves an alternating walking pattern, with the hind print registered on the fore print. When a Woodchuck runs from danger, it leaves groups of four prints, hind prints ahead of fore prints.

Similar Species: A small Raccoon's (p. 46) bounding track pattern will be similar, but will show five-toed fore prints.

Eastern Chipmunk

fore

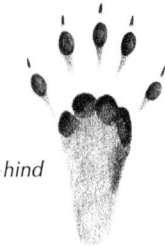

hind

Fore Print
Length: 2–2.5 cm (0.8–1 in)
Width: 1–2 cm (0.4–0.8 in)
Hind Print
Length: 1.8–3.3 cm (0.7–1.3 in)
Width: 1.3–2.3 cm (0.5–0.9 in)
Straddle
5–8 cm (2–3.2 in)
Stride
Bounding: 18–38 cm (7–15 in)
Size
Length with tail: 18–25 cm (7–10 in)
Weight
70–140 g (2.5–5 oz)

bounding

EASTERN CHIPMUNK
Tamias striatus

This large chipmunk is found in a variety of habitats, from the dense forest floor to open areas near buildings. Look for this delightful character in the Maritimes. You are more likely to see or hear this rodent, which is highly active during summer, than to notice its tracks. This chipmunk is happiest on the ground, but it will gladly climb sturdy oak trees to harvest juicy, ripe acorns. Eastern Chipmunks enter a deep sleep in winter, waking up from time to time to have a meal.

Chipmunks are so light that their tracks rarely show fine details. The forefeet each have four toes and the hind feet five. Chipmunks run on their toes, so the two heel pads of the forefeet seldom register; the hind feet have no heel pads. Their erratic track patterns, like those of many of their cousins, show the hind feet registered in front of the forefeet. A chipmunk trail often leads to extensive burrows.

Similar Species: Midwinter tracks are likely a squirrel's (pp. 80–83). Red Squirrel (p. 80) tracks are larger and have a larger straddle. Mouse (pp. 88–91) tracks are smaller.

Red Squirrel

fore

hind

Fore Print
Length: 2–3.8 cm (0.8–1.5 in)
Width: 1.3–2.5 cm (0.5–1 in)

Hind Print
Length: 3.8–5.8 cm (1.5–2.3 in)
Width: 2–3.3 cm (0.8–1.3 in)

Straddle
7.5–11 cm (3–4.5 in)

Stride
Bounding: 20–75 cm (8–30 in)

Size
Length with tail:
    23–38 cm (9–15 in)

Weight
55–260 g (2–9 oz)

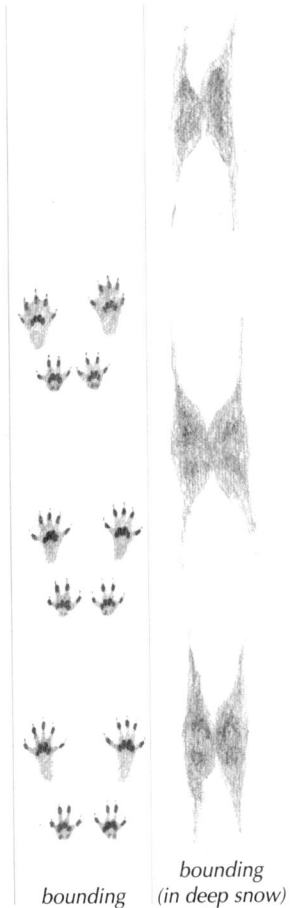

bounding

*bounding
(in deep snow)*

RED SQUIRREL
(Pine Squirrel, Chickaree)
Tamiasciurus hudsonicus

When you enter a
Red Squirrel's territory,
the inhabitant greets you with
a loud, chattering call. Another
sign of this forest dweller, which is found throughout most
of Atlantic Canada, except Newfoundland, is the presence
of middens—large piles of cone scales and cores left beneath
trees—that indicate favourite feeding sites.

Active year-round in their small territories, Red Squirrels
leave an abundance of trails that lead from tree to tree or
down a burrow. These energetic animals mostly bound,
leaving groups of four prints; the hind prints appear in front
of the fore prints, which tend to be side by side (but not
always). Four toes show on each fore print, and five show
on each hind print. The heels often do not register when
squirrels move quickly. In deeper snow the prints merge
to form pairs of diamond-shaped tracks.

Similar Species: Eastern Chipmunks (p. 78) and Northern
Flying Squirrels (p. 82) make tracks in a similar pattern, but
have a narrower straddle and smaller prints.

Northern Flying Squirrel

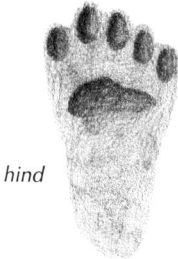

fore

hind

Fore Print
Length: 1.3–2 cm (0.5–0.8 in)
Width: 1.3 cm (0.5 in)

Hind Print
Length: 3.3–4.5 cm (1.3–1.8 in)
Width: 2 cm (0.8 in)

Straddle
7.5–9.5 cm (3–3.8 in)

Stride
Bounding: 28–75 cm (11–30 in)

Size
Length with tail: 23–30 cm (9–12 in)

Weight
110–180 g (4–6.5 oz)

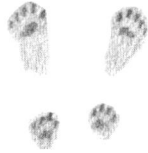

sitzmark into bounding

NORTHERN FLYING SQUIRREL
Glaucomys sabrinus

This acrobat can be found in coniferous forests throughout most of Atlantic Canada, except Newfoundland. It
prefers widely spaced forests, where it can glide from tree to
tree by night. Northern Flying Squirrels will den up together
in a tree cavity for warmth.

Because of its gliding, this squirrel does not leave as
many tracks as other squirrels do. Evidence is scarce in
summer, but in winter you may come across a sitzmark (the
distinctive pattern made when it landed in the snow) and a
short bounding trail, made as it rushed off to the nearest
tree or to do some quick foraging. The bounding track
pattern is typical of squirrels and other rodents, but with the
hind feet registering only slightly in front of the forefeet–
though often all four feet register in a row.

Similar Species: The Red Squirrel (p. 80) usually makes
larger prints and rarely leaves a sitzmark, but unclear tracks
in deep snow can be identical. Eastern Chipmunks (p. 78)
make smaller prints with a narrower straddle.

Norway Rat

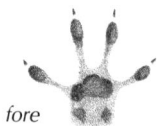

fore

hind

Fore Print
Length: 1.8–2 cm (0.7–0.8 in)
Width: 1.3–1.8 cm (0.5–0.7 in)
Hind Print
Length: 2.5–3.3 cm (1–1.3 in)
Width: 2–2.5 cm (0.8–1 in)
Straddle
5–7.5 cm (2–3 in)
Stride
Walking: 3.8–9 cm (1.5–3.5 in)
Bounding: 23–50 cm (9–20 in)
Size
Length with tail: 33–48 cm (13–19 in)
Weight
200–510 g (7–18 oz)

walking

NORWAY RAT
(Brown Rat)
Rattus norvegicus

Active both day and night, this despised rat is widespread almost anywhere that humans have decided to build their homes. Not entirely dependent on people, it may live in the wild as well.

The fore print shows four toes, and the hind print shows five. When it bounds, this colonial rat leaves four-print groups, with the hind prints in front of the diagonally placed fore prints. Sometimes one of the hind feet direct registers on a fore print, creating a three-print group. This rat more commonly leaves an alternating walking pattern with the larger hind prints close to or overlapping the fore prints; the hind heel does not show. The tail often leaves a dragline in snow. Rats live in groups, so you may find many trails together, often leading to their 2-cm (5-in) wide burrows.

Similar Species: Mouse (pp. 88–91) prints are much smaller. Eastern Chipmunk (p. 78) tracks are similar but smaller, and the gaits differ. Red Squirrel (p. 80) tracks show distinctive squirrel traits.

Meadow Vole

fore

hind

Fore Print
Length: 0.8–1.3 cm (0.3–0.5 in)
Width: 0.8–1.3 cm (0.3–0.5 in)

Hind Print
Length: 0.8–1.5 cm (0.3–0.6 in)
Width: 0.8–1.5 cm (0.3–0.6 in)

Straddle
3.3–5 cm (1.3–2 in)

Stride
Walking/Trotting:
   3.3–7.5 cm (1.3–3 in)
Bounding: 10–20 cm (4–8 in)

Size
Length with tail:
   14–20 cm (5.5–8 in)

Weight
14–70 g (0.5–2.5 oz)

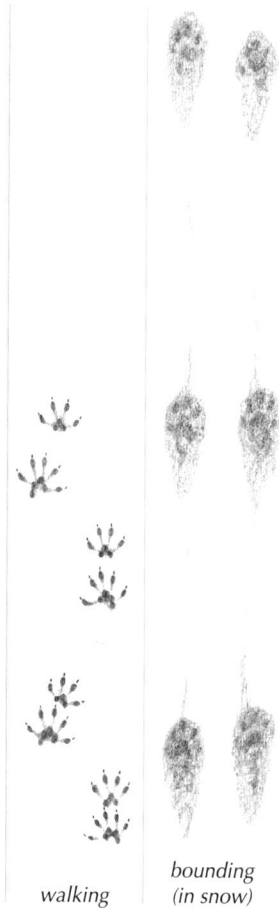

walking

*bounding
(in snow)*

MEADOW VOLE
(Field Mouse)
Microtus pennsylvanicus

With so many vole species in the region, positive track identification is next to impossible, but note that the Meadow Vole frequents lush, damp or wet habitats.

When clear (which is seldom), vole fore prints show four toes, and hind prints show five. A vole's walk and trot both leave a paired alternating track pattern with a hind print occasionally direct registered on a fore print. Voles usually opt for a faster bounding; the resulting print pairs show the hind prints registered on the fore prints. These voles lope quickly across open areas, which creates a three-print track pattern. In winter, voles stay under the snow; when it melts, look for distinctive piles of cut grass from their ground nests. The bark at the bases of shrubs may show tiny teeth marks left by gnawing. In summer, well-used vole paths appear as little runways in the grass.

Similar Species: Other common voles, with similar tracks, include the Southern Red-backed Vole (*Cletherionomys gapperi*), the Heather Vole (*Phenacomys intermedius*) and the Rock Vole (*M. chrotorrhinus*). Deer Mouse (p. 88) bounding tracks show four-print groups.

Deer Mouse

hind

fore

bounding group

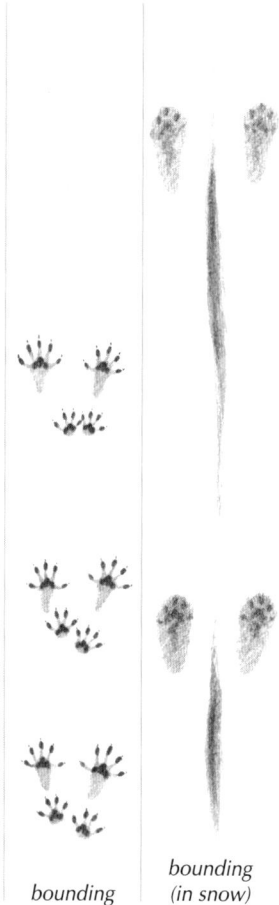

Fore Print
Length: 0.8–1 cm (0.3–0.4 in)
Width: 0.8–1 cm (0.3–0.4 in)

Hind Print
Length: 0.8–1.3 cm (0.3–0.5 in)
Width: 0.8–1 cm (0.3–0.4 in)

Straddle
3.6–4.5 cm (1.4–1.8 in)

Stride
Bounding: 13–30 cm (5–12 in)

Size
Length with tail:
    15–30 cm (6–12 in)

Weight
14–35 g (0.5–1.3 oz)

bounding

*bounding
(in snow)*

DEER MOUSE
Peromyscus maniculatus

The highly adaptable Deer Mouse—one of the most abundant mammals in this region—lives anywhere from arid valleys to alpine meadows. It is seldom seen, because it is nocturnal. The Deer Mouse may enter buildings in winter, where it will stay active.

The fore prints each show four toes, three palm pads and two heel pads. The hind prints show five toes and three palm pads; the heel pads rarely register. It takes perfect, soft mud to get clear prints from such a tiny mammal. Bounding tracks, most noticeable in snow, show the hind prints falling in front of the fore prints. In soft snow the prints may merge to look like larger pairs of prints; tail drag will be evident. A mouse trail may lead up a tree or down into a burrow.

Similar Species: The tracks of many less common species of mice are identical. House Mouse (*Mus musculus*) tracks are very similar, but they are associated more with humans. Voles (p. 86) tend to trot and have a much shorter bounding track pattern. Jumping mouse (p. 90) prints may be similar in size, but show long, thin toes. Chipmunks (p. 78) have a wider straddle. Shrews (p. 92) have a narrower straddle.

Meadow Jumping Mouse

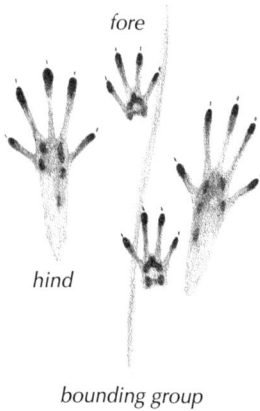

fore

hind

bounding group

Fore Print
Length: 0.8–1.3 cm (0.3–0.5 in)
Width: 0.8–1.3 cm (0.3–0.5 in)
Hind Print
Length: 1.3–3.3 cm (0.5–1.3 in)
Width: 1.3–1.8 cm (0.5–0.7 in)
Straddle
4.5–4.8 cm (1.8–1.9 in)
Stride
Bounding: 18–45 cm (7–18 in)
In alarm: 90–180 cm (3–6 ft)
Size
Length with tail: 18–23 cm (7–9 in)
Weight
17–35 g (0.6–1.3 oz)

bounding

MEADOW JUMPING MOUSE
Zapus hudsonius

Congratulations
if you find and successfully
identify the tracks of the
Meadow Jumping Mouse!
Though it lives throughout most of the region, its preference
for grassy meadows and its long, deep winter hibernation
(about six months!) make locating tracks very difficult.

Jumping mouse tracks are distinctive if you do find them.
The two smaller forefeet register between the long hind feet;
the long heels do not always register and some prints show
just the three long middle toes. The toes on the forefeet may
splay so much that the side toes point backward. When they
bound, jumping mice make short leaps. The tail may leave
a dragline in soft mud or unseasonable snow. Clusters of cut
grass stems about 13 cm (5 in) long and lying in meadows
are a more abundant sign of this rodent.

Similar Species: The Woodland Jumping Mouse (*Napaeo-zapus insignis*), with similar tracks, is found in some areas.
Deer Mouse (p. 88) tracks may have the same straddle.
Heel-less hind prints may be mistaken for a vole's (p. 86)—
or a small bird's (p. 116) or an amphibian's (pp. 118–123).

Masked Shrew

hind

fore

bounding group

bounding

Fore Print
Length: 0.5 cm (0.2 in)
Width: 0.5 cm (0.2 in)

Hind Print
Length: 1.5 cm (0.6 in)
Width: 0.8 cm (0.3 in)

Straddle
2–3.3 cm (0.8–1.3 in)

Stride
Bounding: 3–7.5 cm (1.2–3 in)

Size
Length with tail: 7–11 cm (2.3–4.5 in)

Weight
3–9 g (0.1–0.3 oz)

MASKED SHREW

Sorex cinereus

Though several species of tiny, frenetic shrews are found in the region, the widespread and adaptable Masked Shrew is a likely candidate if you find tracks. This small shrew prefers moist fields, marshes, bogs or woodlands, but it can be found in higher, drier grasslands. Its rapid activity makes it difficult to observe closely.

In its energetic and unending quest for food, a shrew usually leaves a four-print bounding pattern, but it may slow to an alternating walking gait. The individual prints in a group are often indistinct, but in mud or shallow, wet snow you can even count the five toes on each print. In snow, a shrew's tail often leaves a dragline. If a shrew tunnels under snow, it may leave a distinct snow ridge on the surface. A shrew's trail may disappear down a burrow.

Similar Species: Other abundant shrew species with similar tracks include the small Pygmy Shrew (*Sorex hoyi*), which is also common in moist wooded areas, the widespread Northern Short-tailed Shrew (*Blarina brevicauda*), the Smoky Shrew (*Sorex fumeus*) and the Gaspé Shrew (*Sorex gaspensis*). Mouse (pp. 88–91) fore prints show four toes.

Star-nosed Mole

a molehill of the Star-nosed Mole

Size
Length with tail: 15–22 cm (6–8.5 in)
Weight
28–75 g (1–2.6 oz)

STAR-NOSED MOLE
Condylura cristata

This peculiar character is found throughout the region, except in Newfoundland. The Star-nosed Mole is easily identified by the strange tentacle-like protrusions on its nose, thought to help it find food in its dark, subterranean world. It spends more time out of its burrow than other moles do.

Typical signs of this mole include the big piles of mud pushed out of its burrows. Because of its preference for swimming and wet areas, look for this mole's hills along the banks of streams and rivers and in raised areas around marshes and wet fields. Clear prints from its long-clawed feet are rarely found. The Star-nosed Mole remains under the snow during winter and swims under the ice, so winter tracks are seldom seen.

Similar Species: There are no other moles in the region, and no other small animal leaves similar signs.

BIRDS, AMPHIBIANS & REPTILES

A guide to the animal tracks of Atlantic Canada is not complete without some consideration of the many birds and amphibians found in the region.

Several bird species have been chosen to represent the main types common to this region. Remember, however, that individual bird species are not easily identified by track alone. Bird tracks can often be found in abundance in snow and are clearest in shallow, wet snow. The shores of streams and lakes are very reliable locations in which to find bird tracks—the mud there can hold a clear print for a long time. The sheer number of tracks made by shorebirds and water-fowl can be astonishing. Though some bird species prefer to perch in trees or soar across the sky, it can be quite entertaining to follow the tracks of birds that spend a lot of time on the ground. They can spin around in circles and lead you in all directions. The trail may suddenly end as the bird takes flight, or it might terminate in a pile of feathers, the bird having fallen victim to a hungry predator.

Many amphibians and turtles depend on moist environments, so look in the soft mud along the shores of lakes and ponds for their distinctive tracks. Though you may be able to distinguish frog tracks from toad tracks, because they generally move differently, it can be very difficult to identify the species. In drier environments, reptiles, which thrive in dryness, outnumber the amphibians. Unfortunately, because of this preference for dry terrain, most reptiles seldom leave good tracks for us to identify.

Canada Goose

Print
Length: 10–13 cm (4–5 in)
Straddle
13–18 cm (5–7 in)
Stride
Walking: 13–18 cm (5–7 in)
Size
80–120 cm (2.7–4 ft)

CANADA GOOSE
Branta canadensis

This common goose is a familiar sight in open areas by lakes and ponds. Its huge, webbed feet leave prints that can often be seen in abundance along the muddy shores of just about any waterbody, including those in urban parks, where the Canada Goose's green-and-white droppings often accumulate in prolific amounts.

The webbed feet each have three long toes, all facing forward. These toes register well, but the webbing between them does not always show on the print. The feet point inward, giving the prints a pigeon-toed appearance and perhaps accounting for the bird's waddling gait.

Similar Species: Many waterfowl, as well as gulls (p. 100), leave similar prints.

Herring Gull

Print
Length: 9 cm (3.5 in)
Straddle
10–15 cm (4–6 in)
Stride
11 cm (4.5 in)
Size
Length: 58–65 cm (23–25 in)

HERRING GULL

Larus argentatus

The Herring Gull, with its long wings and webbed toes, is a strong long-distance flier and an excellent swimmer. It is common in a variety of habitats and is concentrated in great numbers on the coast.

Gulls leave slightly asymmetrical tracks that show three toes. They have claws that register outside the webbing, and the claw marks are usually attached to the footprint. Most gulls have quite a swagger to their gait, and they leave a trail with the tracks turned strongly inward.

Similar Species: Gull species cannot be reliably identified by track alone, but smaller species have conspicuously smaller tracks. Duck tracks are often difficult to distinguish from gull tracks.

Great Blue Heron

Print
Length: to 17 cm (6.5 in)
Straddle
20 cm (8 in)
Stride
23 cm (9 in)
Size
1.3–1.4 m (4.2–4.5 ft)

GREAT BLUE HERON
Ardea herodias

The refined and graceful image of this large heron has come to symbolize the precious wetlands in which it patiently hunts for food. Usually still and statuesque as it waits for a meal to swim by, this heron will have cause to walk from time to time, perhaps to find a better hunting location. Look for its large, slender tracks along the banks or mudflats of waterbodies.

Not surprisingly, a bird that lives and hunts with such precision walks in a similar fashion, leaving straight tracks that fall in a nearly straight line. Look for the slender rear toe in the print.

Similar Species: Cranes (*Grus* spp.), which occupy similar habitats, have similar and possibly larger tracks, but a crane's rear toes are smaller and do not register.

Common Snipe

Print
Length: 3.8 cm (1.5 in)
Straddle
to 4.5 cm (1.8 in)
Stride
to 3.3 cm (1.3 in)
Size
28–30 cm (11–12 in)

COMMON SNIPE

Gallinago gallinago

This short-legged character is a resident of marshes and bogs, where its neat prints can often be seen in mud. Snipes are quite secretive when on the ground, so you may be surprised if one suddenly flushes out from beneath your feet. If there is a Common Snipe in the air, you may hear an eerie whistle as it dives from the sky.

The Common Snipe's neat prints show four toes, including a small rear toe that points inward. The bird's short legs and stocky body give it a very short stride.

Similar Species: Many shorebirds, including the Spotted Sandpiper (p. 106), leave similar tracks.

Spotted Sandpiper

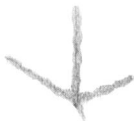

Print
Length: 2–3.3 cm (0.8–1.3 in)
Straddle
to 3.8 cm (1.5 in)
Stride
Erratic
Size
18–20 cm (7–8 in)

SPOTTED SANDPIPER
Actitis macularia

The bobbing tail of the Spotted Sandpiper is a common sight on the shores of lakes, rivers and streams, but you will usually find just one of these territorial birds in any given location. Because of its excellent camouflage, likely the first that you will see of this bird will be when it flies away, its fluttering wings close to the surface of the water.

As it teeters up and down on the shore, a sandpaper leaves trails of three-toed prints. Its fourth toe is very small and faces off to one side at an angle. Sandpiper tracks can have an erratic stride.

Similar Species: All sandpipers and plovers, including the common Killdeer (*Charadrius vociferus*), leave similar tracks, although there is much diversity in size. The Common Snipe (p. 104) makes larger tracks.

Ruffed Grouse

Print
Length: 5–7.5 cm (2–3 in)
Straddle
5–7.5 cm (2–3 in)
Stride
Walking: 7.5–15 cm (3–6 in)
Size
38–48 cm (15–19 in)

RUFFED GROUSE
Bonasa umbellus

   This ground-dweller prefers the quiet seclusion of coniferous forests in winter, so that will be the best place to find its tracks. If you follow them quietly, you may be startled when the grouse bursts from cover underneath your feet. Its excellent camouflage usually affords it good protection.

   The three thick front toes leave very clear impressions, but the short rear toe, which is angled off to one side, does not always show up so well. This bird's neat, straight trail appears to reflect its conservative and cautious approach to life on the forest floor.

Similar Species: Other grouse leave similar tracks, but their prints may be blurred and enlarged by the winter feathers that grow on their feet.

Great Horned Owl

Strike
Width: to 90 cm (3 ft)
Size
55 cm (22 in)

GREAT HORNED OWL
Bubo virginianus

Often seen resting quietly in trees by day, this wide-ranging owl prefers to hunt at night. An accomplished hunter in snow, the owl strikes through the snow with its talons, leaving an untidy hole that may be surrounded by wing and tail-feather imprints. If it registers well, this 'strike' can be quite a sight. The feather imprints are made as the owl struggles to take off with possibly heavy prey. An ungraceful walker, it prefers to fly away from the scene.

You may stumble across a strike and guess that the owl's target could have been a vole (p. 86) scurrying around underneath the snow. Or you may be following the surface trail of an animal, only to find that it abruptly ends with this strike mark, where the animal has been seized by an owl, or perhaps a hawk or raven.

Similar Species: If the prey left no approaching trail, the strike mark is likely an owl's, because owls hunt by sound. If there is a trail, the strike mark, usually with less rounded and more distinct feather imprints, could be that of a hawk or a Common Raven (p. 112), both of whom hunt by sight.

111

Common Raven

Print
Length: to 10 cm (4 cm)

Straddle
to 10 cm (4 cm)

Stride
Walking: to 15 cm (6 cm)

Size
60 cm (2 ft)

COMMON RAVEN
Corvus corax

   This legendary bird spends a lot of time strutting around on the ground—confident behaviour that may hint at its considerable intelligence.

   Ravens leave a typical alternating track pattern. Prints show three thick toes pointing forward and one toe pointing backward. When a Raven is in need of greater speed, such as for takeoff, it leaves a trail of diagonally placed pairs of prints that are rather irregular.

Similar Species: Other corvids, such as the American Crow (p. 114), also spend a lot of time poking around on the ground. Their tracks are similar, but smaller, and their strides are correspondingly shorter.

American Crow

Print
Length: 6.5–7.5 cm (2.5–3 in)
Straddle
3.8–7.5 cm (1.5–3 in)
Stride
Walking: 10 cm (4 in)
Size
40 cm (16 in)

AMERICAN CROW
Corvus brachyrhyncos

The black silhouette of the American Crow can be a common sight in a variety of habitats. Like the Common Raven (p. 112), a crow will frequently come down to the ground and contentedly strut around with a confidence that hints at its intelligence. Its loud *caw* can be heard from quite a distance; crows can be especially noisy when they are mobbing an owl or hawk.

The American Crow typically leaves an alternating walking track pattern. Its prints show three sturdy toes pointing forward and one toe pointing backward. When a crow is in need of greater speed, perhaps for takeoff, it bounds along, leaving irregular pairs of diagonally placed prints with a longer stride between each pair.

Similar Species: Other corvids, such as jays, also spend a lot of time on the ground and make similar tracks. The much larger Common Raven leaves larger tracks.

Dark-eyed Junco

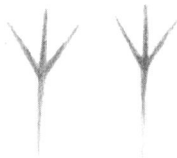

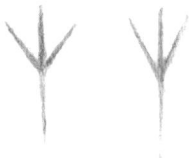

Print
Length: to 3.8 cm (1.5 in)
Straddle
2.5–3.8 cm (1–1.5 in)
Stride
Hopping: 3.8–13 cm (1.5–5 in)
Size
14–17 cm (5.5–6.5 in)

DARK-EYED JUNCO
Junco hyemalis

This common small bird typifies the many small hopping birds found in the region. Each foot has three forward-pointing toes and one longer toe at the rear. The best prints are left in snow, although in deep snow the toe detail is lost; the footprints may show some dragging between the hops.

A good place to study this type of prints is near a bird-feeder. Watch the birds scurry around as they pick up fallen seeds, then have a look at the prints left behind. For example, juncos are attracted to seeds that chickadees (*Poecile* spp.) scatter as they forage for sunflower seeds in the bird-feeder. Also look for tracks under coniferous trees, where juncos feed on fallen seeds in winter.

Similar Species: Toe size may help with identification—larger birds make larger prints—as can the season. In powdery snow, junco tracks could be confused with mouse (pp. 88–91) tracks, so follow the trail to see if it disappears down a hole or into thin air.

Frogs

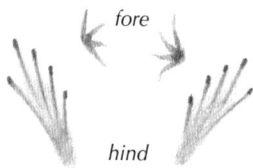

fore

hind

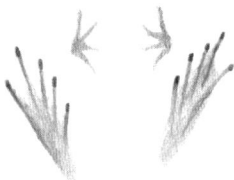

Straddle
to 7.5 cm (3 in)

hopping

FROGS

Wood Frog

The best place to look for frog tracks is along the muddy fringes of waterbodies.

The smallest frogs include the treefrogs. The Spring Peeper's (*Pseudoacris crucifer*) 3.8-cm (1.5-in) length and its preference for thick undergrowth and shrubs near the water make its tracks a rare sight. The widespread Wood Frog (*Rana sylvatica*), often found in dry woodland areas seemingly far from water, is larger, at about 9 cm (3.5 in) long. The Green Frog (*R. clamitans*) and Pickerel Frog (*R. palustris*) are also widespread. The beautiful and wide-spread Northern Leopard Frog (*R. pipiens*), to 13 cm (5 in) long, makes larger tracks. Unusually large tracks are surely from the robust Bullfrog (*R. catesbeiana*). Growing to 20 cm (8 in) in length, it is North America's largest frog.

A frog's hopping action results in its two small forefeet registering in front of its long-toed hind prints. Frog tracks vary greatly in size, depending on species and age. Toads (p. 120) may also hop, but they usually walk.

Toads

hind   *fore*

Straddle
to 6.5 cm (2.5 in)

walking

TOADS

American Toad

In Atlantic Canada there are fewer toad species than frog species. The best place to look for toad tracks is, as with frogs, undoubtedly along the muddy fringes of waterbodies, but their tracks can also be found in drier areas, as unclear trails in dusty patches of soil.

The toad most likely to be encountered, and the most widespread, is the American Toad (*Bufo americanus*), which lives in many different moist habitats. Toads in this region can be up to 4.5 inches (11 cm) in length.

In general, toads walk and frogs (p. 118) hop, but toads are pretty capable hoppers, too, especially when they are hassled by overly enthusiastic naturalists. Toads leave rather abstract prints as they walk. The heels of the hind feet do not register. On less firm surfaces, the toes often leave draglines.

Salamanders & Newts

fore

hind

Straddle
to 7.5 cm (3 in)

walking

SALAMANDERS & NEWTS

*Redback
Salamander*

There are several kinds of salamanders and newts in the moist and wet areas of the southern parts of Atlantic Canada. Among the more abundant and widespread of these long, slender, lizard-like amphibians is the Eastern Newt (*Notophthalmus viridescens*), which can grow to 14 cm (5.5 in) in length. After a fresh rain, Eastern Newts emerging from ponds may leave small trails in the mud.

Other species of salamanders outnumber the newts in this region. The Redback Salamander (*Plethodon cinereus*), which can reach 13 cm (5 in) in length, lives throughout the region in mixed forests and in some coniferous forests. The beautiful Blue-spotted Salamander (*Ambystoma laterale*), which grows to be about the same length, favours moist lowlands of deciduous forests.

In general, a salamander or newt fore print shows four toes, and the larger hind print shows five. However, print detail is often blurred by the animal's dragging belly or by the swinging of its thick tail across the tracks.

Turtles

fore

hind

fore

hind

Straddle
10–25 cm (4–10 in)

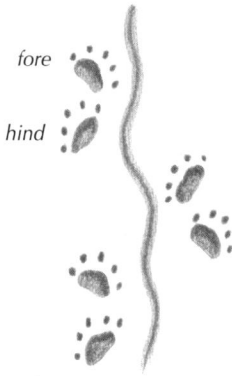

Snapping Turtle walking

typical turtle walking

TURTLES

Wood Turtle

Turtles, those ancient inhabitants of the water world, will happily slip into the murky depths of the water to avoid detection. They do, however, come out from time to time to feed or bask in the sunshine. Look for their distinctive tracks alongside ponds, rivers and moist areas. Note that some turtles, such as the huge Snapping Turtle (*Chelydra serpentina*), prefer to stay in the water and rarely come out.

One of the most widespread and prettiest turtles, which is often seen basking, is the Painted Turtle (*Chrysemys picta*); it can reach almost 25 cm (10 in) in length. Slightly smaller is the Wood Turtle (*Clemmys insculpta*), which likes moist forested areas and woodland streams. Both of these turtles are found in southern parts of Atlantic Canada.

With its large shell and short legs, a turtle leaves a track that is wide relative to the length of its stride—its straddle is about half its body length. Although longer-legged turtles can raise their shells off the ground, short-legged species may let their shells drag, as shown in their tracks. The tail may leave a straight dragline in the soft mud. On firmer surfaces, look for distinct claw marks.

Snakes

typical snake track

SNAKES

*Common
Garter Snake*

There are several species of snakes to be found in Atlantic Canada, with greater diversity in the warmer south. Because snakes are all long and slender, their tracks are so similar that identification among the species is next to impossible. Because a snake lacks feet and leaves a track that is just a gentle meander, it can be quite a challenge even to establish in which direction the snake was moving.

The most widespread and frequently encountered snake is the Common Garter Snake (*Thamnophis sirtalis*). Found throughout the region, often close to wet or moist areas, this harmless snake can reach 1.3 m (4.3 ft) in length.

Other widespread species include the Red-bellied Snake (*Storeria occipitomaculata*), which prefers hilly woodlands and reaches a length of up to 40 cm (16 in). Only slightly larger, the Smooth Green Snake (*Opheodrys vernalis*) inhabits grassy meadows and fields along forest edges. The elusive Ring-neck Snake (*Diadophis punctatus*) prefers to hide under clumps of leaves, grass or forest debris. This secretive snake lives in a fairly wide variety of habitats, from grasslands to deciduous forests.

TRACK PATTERNS & PRINTS

Coyote
p. 30

Gray Wolf
p. 32

Red Fox
p. 34

Arctic Fox
p. 36

Domestic Dog
p. 38

Lynx
p. 40

TRACK PATTERNS & PRINTS

Bobcat
p. 42

Domestic Cat
p. 44

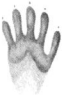

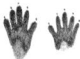

Raccoon
p. 46

Harbour Seal
p. 48

River Otter
p. 50

Fisher
p. 52

130

Marten
p. 54

Mink
p. 56

Least Weasel
p. 58

Short-tailed Weasel
p. 60

Long-tailed Weasel
p. 62

Wolverine
p. 64

131

TRACK PATTERNS & PRINTS

Striped Skunk
p. 66

Snowshoe Hare
p. 68

Porcupine
p. 70

Beaver
p. 72

Muskrat
p. 74

Woodchuck
p. 76

132

Eastern Chipmunk
p. 78

Red Squirrel
p. 80

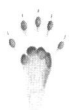

Northern
Flying Squirrel
p. 82

Norway Rat
p. 84

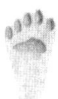

Meadow Vole
p. 86

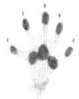

Deer Mouse
p. 88

133

TRACK PATTERNS & PRINTS

Meadow
Jumping Mouse
p. 90

Masked Shrew
p. 92

Canada Goose
p. 98

Herring Gull
p. 100

Great Blue Heron
p. 102

Common Snipe
p. 104

Spotted Sandpiper
p. 106

Ruffed Grouse
p. 108

Common Raven
p. 112

American Crow
p. 114

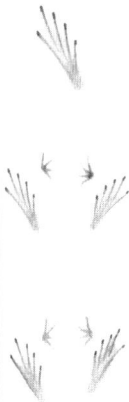

Dark-eyed Junco
p. 116

Frogs
p. 118

135

Toads
p. 120

Salamanders
p. 122

Turtles
p. 124

Snakes
p. 126

HOOFED PRINTS

White-tailed
Deer

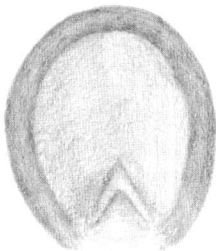

Horse

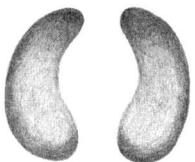

Caribou

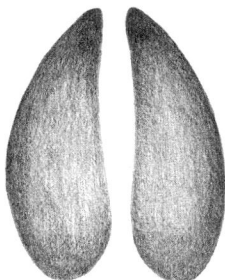

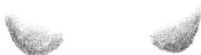

Moose

inch   cm
0 ┬ 0
  ┤
1 ┤
  ┤
2 ┴ 5

137

HIND PRINTS

Deer
Mouse

Meadow
Vole

Masked
Shrew

Northern
Flying
Squirrel

Red
Squirrel

Woodchuck

Muskrat

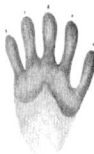

Raccoon

Porcupine

inch cm

0 — 0

— 1

— 2

1 — 3

— 4

2 — 5

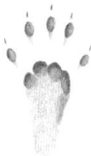

Eastern
Chipmunk

Meadow
Jumping Mouse

Norway
Rat

138

HIND PRINTS

Snowshoe Hare

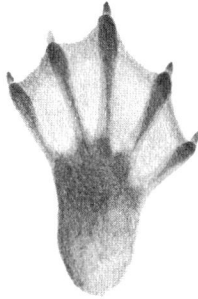

Beaver

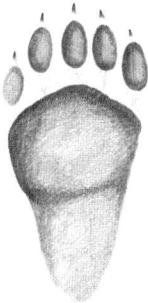

Black Bear

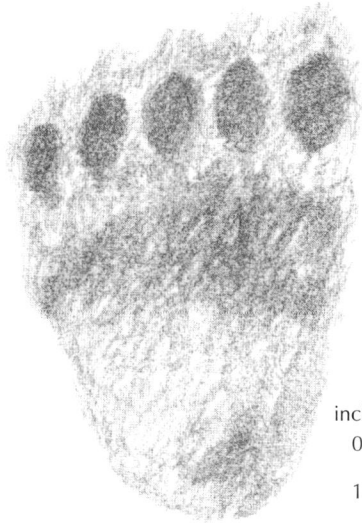

Polar Bear

inch cm
0 — 0
1 —
2 — 5

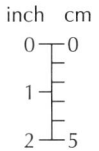

139

FORE PRINTS

Least
Weasel

Short-tailed
Weasel

Long-tailed
Weasel

Mink

Domestic
Cat

Coyote

Bobcat

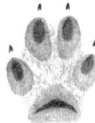

Red Fox

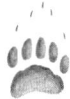

Striped Skunk

Fisher

Arctic Fox

Marten

River Otter

inch cm

0 — 0

1

2 — 5

Lynx

Wolverine

BIBLIOGRAPHY

Behler, J.L., and F.W. King. 1979. *Field Guide to North American Reptiles and Amphibians*. National Audubon Society. New York: Alfred A. Knopf.

Brown, R., J. Ferguson, M. Lawrence and D. Lees. 1987. *Tracks and Signs of the Birds of Britain and Europe: An Identification Guide*. London: Christopher Helm.

Burt, W.H. 1976. *A Field Guide to the Mammals*. Boston: Houghton Mifflin Company.

Farrand, J., Jr. 1995. *Familiar Animal Tracks of North America*. National Audubon Society Pocket Guide. New York: Alfred A. Knopf.

Forrest, L.R. 1988. *Field Guide to Tracking Animals in Snow*. Harrisburg: Stackpole Books.

Halfpenny, J. 1986. *A Field Guide to Mammal Tracking in North America*. Boulder: Johnson Publishing Company.

Headstrom, R. 1971. *Identifying Animal Tracks*. Toronto: General Publishing Company.

Murie, O.J. 1974. *A Field Guide to Animal Tracks*. The Peterson Field Guide Series. Boston: Houghton Mifflin Company.

Rezendes, P. 1992. *Tracking and the Art of Seeing: How to Read Animal Tracks and Sign*. Vermont: Camden House Publishing.

Stall, C. 1989. *Animal Tracks of the Rocky Mountains*. Seattle: The Mountaineers.

Stokes, D., and L. Stokes. 1986. *A Guide to Animal Tracking and Behaviour*. Toronto: Little, Brown and Company.

Wassink, J.L. 1993. *Mammals of the Central Rockies*. Missoula: Mountain Press Publishing Company.

Whitaker, J.O., Jr. 1996. *National Audubon Society Field Guide to North American Mammals*. New York: Alfred A. Knopf.

INDEX